Materials

Rock

Chris Oxlade

www.heinemann.co.uk/library
Visit our website to find out more information about **Heinemann Library** books.

To Order:

☎ Phone 44 (0) 1865 888066

▤ Send a fax to 44 (0) 1865 314091

▭ Visit the Heinemann Library Bookshop at www.heinemann.co.uk/library to browse our catalogue and order online.

First published in Great Britain by Heinemann Library, Halley Court, Jordan Hill, Oxford OX2 8EJ a division of Reed Educational and Professional Publishing Ltd.
Heinemann is a registered trademark of Reed Educational & Professional Publishing Ltd.

OXFORD MELBOURNE AUCKLAND JOHANNESBURG BLANTYRE
GABORONE IBADAN PORTSMOUTH (NH) USA CHICAGO

Designed by Storeybooks
Originated by Ambassador Litho Ltd.
Printed and bound in Hong Kong/China

ISBN 0 431 12738 7 (hardback)
06 05 04 03 02
10 9 8 7 6 5 4 3 2

ISBN 0 431 12744 1 (
06 05 04 03 02
10 9 8 7 6 5 4 3 2 1

British Library Cataloguing in Publication Data
    Oxlade, Chris
        Rock. – (Materials)
        1.Rocks – Juvenile literature
        I.Title
        552

Acknowledgements
The Publishers would like to thank the following for permission to reproduce photographs:
Corbis: pp10, 15, 17; David Kampfner: p20; Ecoscene: p14, Andrew Brown p4, Anthony Cooper p25, Tony Page p27, Barry Webb p16, Winkley p6; Eye Ubiquitous: p21; FLPA: p12; GSF Picture Library: pp8, 9, 18, 24, © Dr B Booth p7; Impact: Piers Cavendish p19; Martyn Chillmaid: pp5, 29; Oxford Scientific Films: Stuart Bebb p26, Harold Taylor Abipp p13; Trip: H Rogers p22; Tudor Photography: pp11, 23.

Cover photograph reproduced with permission of Tudor Photography.

Every effort has been made to contact copyright holders of any material reproduced in this book. Any omissions will be rectified in subsequent printings if notice is given to the Publisher.

Contents

What is rock?   4

Where rocks come from   6

More types of rock   8

Hard and soft   10

Colour and pattern   12

Weathering   14

Finding rocks   16

Building with rock   18

Rocks on the ground   20

Rock decoration   22

Making rock shapes   24

Rocks and pollution   26

Fact file   28

Would you believe it?   29

Glossary   30

More books to read   32

Index   32

You can find words shown in bold, **like this,** in the Glossary.

What is rock?

Rock is a **natural** material. There is a thick layer of rock all over the Earth's surface, even under the sea. It is called the Earth's **crust**. You can see the rock in hills and mountains.

Rock is a useful material. We use rock for building homes and roads. We also use it for decorating floors and walls, and for making jewellery.

Where rocks come from

This rock was made from sand that piled up on the bottom of the sea millions of years ago. The sand was buried and pressed together to make the rock. This is called **sedimentary** rock.

This sedimentary rock is called limestone. It was made from sea creatures. When the creatures died, they fell to the bottom of the sea. Their skeletons were pressed together to make limestone.

More types of rock

The Earth's **crust** is made of solid
rock. Underneath the crust is hot,
runny rock called lava. Sometimes lava
leaks through the crust. When it cools
down, lava turns into **igneous** rock.

Sedimentary or igneous rocks are gradually buried under more new rock. Deep underground, heat turns them into another type of rock, called **metamorphic** rock.

Hard and soft

There are many different types of rock. Many rocks are very hard. This kitchen worktop is made from a hard rock called granite.

Some **sedimentary** rocks are soft. Some are so soft that you can break them up with your hands. Chalk is a soft, white rock. It is used for writing.

Colour and pattern

All rocks are made up of materials called **minerals**. Some minerals are black, some are white, pink or green. The colour of a rock depends on the minerals in it.

Rocks have different patterns of colour in them. Some have stripes and some have curls. Marble has beautiful patterns of colour.

Weathering

The rocks on the Earth's surface are battered by rain and wind. They are also heated by the Sun, which makes them crack. Slowly they break into small pieces. This is called **weathering**.

Small pieces of rock are washed into streams and rivers. They bump along in the water. As they do, they wear away the river bed. Over millions of years, this forms deep valleys and giant **canyons**.

Finding rocks

The rocks we use must be dug out of
the ground. This happens in places
called **quarries**. Sometimes the rock is
broken up with **explosives**. Then
diggers scoop up the smashed rock.

Gravel is made up of small, round pieces of rock. We get gravel from places called gravel pits. Huge machines scoop out the gravel and put it into trucks.

Building with rock

Most of the rock we dig from the ground is used for building. We use hard, strong rock for building garden walls and the walls of houses.

These workers are building with a material called **concrete**. It is made from **gravel**, sand and **cement**. When the concrete dries it is as hard as rock.

Rocks on the ground

Hard rock does not wear away easily. It is good for making floors that lots of people walk over. In hot countries floors are often made of stone. The stone helps to keep the room cool.

When you are in a car, you are rolling along on rock. Many road surfaces are made from a material called tarmac. It is a mixture of small bits of hard rock and sticky **tar** that glues the rock together.

Rock decoration

We use rocks with interesting colours and patterns for decoration. This building is covered in a rock called granite. The rock has been polished to make it shiny.

Shiny pieces of rock are also used to make ornaments and jewellery. Sometimes the pieces of rock are cut from larger rocks. Sometimes they are pebbles from a beach.

Making rock shapes

Artists called sculptors use rock to make **sculptures**. They start with a block of rock. They cut bits off with sharp tools until the rock is an interesting shape.

Pieces of rock are also cut into shapes to decorate buildings. The people who cut the rock are called **stonemasons**. They often replace parts of old stone buildings with pieces of new stone.

Rocks and pollution

Taking rocks from the ground can make a mess. Huge **quarries** in the countryside look ugly. Trucks from quarries make lots of noise and dirt.

Pollution from cars and factories mixes up with rain. When the rain falls on rocks, it slowly eats them away. This **sculpture** has been spoiled by pollution.

Fact file

- Rock is a **natural** material.

- Some types of rock, such as granite, are very hard.

- Some types of rock, such as chalk, are soft.

- Rocks come in many different colours.

- Rocks have different patterns of colour in them.

- Rock does not burn when it is heated.

- Rock does not let electricity flow through it.

- Rock is not attracted by magnets.

Would you believe it?

Pumice is a very special type of rock that comes from volcanoes. It is full of air bubbles, which makes it very light. Pumice is the only rock that floats!

Glossary

canyon   deep valley with steep, rocky sides

cement   thick, gooey material that goes hard a few hours after being made

concrete   material made from gravel, sand and cement. It is as strong as rock and used for building.

crust   thick layer of rock on Earth's surface

explosive   something that makes a lot of energy which breaks something else up

gravel   small pieces of rock

igneous   type of rock made when hot, runny lava cools down and sets

metamorphic   type of rock made when other rocks get very hot under the ground

minerals   materials rocks are made from

natural comes from plants, animals or the rocks in the Earth

pollution rubbish or poisonous chemicals that are thrown on to the ground, or into the air, rivers and seas

quarry place where rocks are dug from the ground

sculpture work of art carved from rock or wood, or made of metal

sedimentary type of rock made when mud, sand or sea creatures fall to the bottom of the sea

stonemason someone who cuts rocks into shapes for building

tar gooey, black liquid that is made from oil

weathering wearing away rocks. Rocks can be weathered by the Sun, wind or rain.

More books to read

Images: Materials and Their Properties
Big Book Compilation
Heinemann Library, 1999

My World of Science
Angela Royston
Heinemann Library, 2001

New Star Science: Materials and Their Uses
Ginn, 2001

New Star Science: Rocks and Soil
Ginn, 2001

Science Files: Rocks and Minerals
Steve Parker
Heinemann Library, 2001

Index

chalk   11, 28
colour of rocks   12–13, 28
Earth's crust   4, 8
granite   10, 22
gravel   17, 19
igneous rocks   8
jewellery   5, 23
lava   8
limestone   7

marble   13
metamorphic rocks   9
pollution   27
quarries   16, 26
road surfaces   21
sculptures   24, 27
sedimentary rocks   6–7, 9, 11
weathering   14

Titles in the *Materials* series include:

Hardback   0 431 12737 9

Hardback   0 431 12738 7

Hardback   0 431 12736 0

Hardback   0 431 12735 2

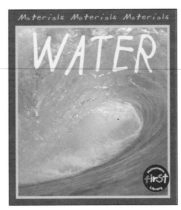

Hardback   0 431 12734 4

Find out about the other titles in this series on our website www.heinemann.co.uk/library